W9-AWG-950

CONTENTS

WHAT IS CYBER SECURITY?

You protect your home and automobile by locking the doors, by not allowing strangers to come inside, and perhaps even with an alarm system. You protect your money by keeping most of it in the bank, and by not sharing your bank card PINs with anyone. It's just as important to protect your personal information—and even your identity—by taking precautions when using computers and the internet.

Cyber security is the protection of computer systems, both hardware (the equipment you use) and software (computer programs), from damage, information theft, or other forms of fraud.

You may store some of your most delicate information on your computer, and you need to make sure you take precautions to keep it secure. Today, computers are more vulnerable to attack than ever before.

Keeping your computer and online information secure can seem like a daunting task at a time when we often hear about hacking, credit card theft, high-profile security leaks, and computer fraud of all kinds.

However, taking the right actions to protect your computer and your personal information does not have to be difficult. For the value and peace of mind it can provide, a few simple steps can provide a substantial return on a very small investment of time.

In this book, we will walk through some of the ways some bad-intentioned people can try to get their hands on your personal information, or trick you into giving them access to items they should not have access to. We will also show you how—by taking a few simple precautions—you can protect yourself, your computer, and your information.

Hacking is motivated by a plethora of reasons. Some hacking campaigns may be motivated by political beliefs or activism, while others are malicious for no apparent reason other than economic gain.

Just as you lock your home and take a close look at those you might decide to open your door to, it's important that you keep your computer and your information safe, and not allow the ill-intentioned easy ways to come inside.

WHAT IS PHISHING?

Phishing can be a remarkably effective way for hackers to lure victims into disclosing their personal information. It typically involves a scammer sending an email that appears to be coming from a legitimate source such as a bank or credit card company. There is usually a tone of urgency to the message. For example, an alert that your account has been hacked or compromised, and a need for you to immediately click a link and type in your user ID and password to verify your ownership of the account. Because the look of the email appears legitimate, it's easy to see why people might click the link and try to login to what they think will be their account information. In reality, they may be clicking to a bogus website, and when they type their ID and password they are going directly to the "phishers." If you're not careful, your password to a bank or credit card account can easily fall into the hands of a scammer.

The reason this form of online scam was dubbed phishing has to do with its similarities to fishing, the sport. Scammers use bait to try to lure unsuspecting people into providing valuable personal information like account numbers, IDs and passwords, and social security numbers.

TYPES OF PHISHING

DECEPTIVE PHISHING This type of phishing involves a deceptive email message designed to trick the user into voluntarily giving up important information.

SPEAR PHISHING This type of phishing involves a high level of personalization, designed to make a user think there is a connection between himself or herself and the sender. For example, the email might include name, title, phone number, or other personal information for a very targeted attack.

WHALING Sometimes called CEO phishing, these attacks are directed toward high-ranking business professionals. They may appear to be a legal issue or important customer complaint that has to be dealt with immediately, with the goal being the attainment of important business information.

DOCUMENT PHISHING An email might appear to come from a service that stores documents, such as Dropbox or Google Drive. The goal could be to lure the user into giving up their login credentials to the service, which the phisher could then access to gain personal information.

HOW TO PREVENT PHISHING

Just as you would take a close look and perhaps ask a few questions before opening your home's door to someone, it's a good idea to give a serious look to any email, instant message, or correspondence (even a phone call) that directs or asks you to provide personal information or account information.

TIPS TO PROTECT YOURSELF

The Federal Trade Commission offers some great suggestions for how to thwart phishing attacks, and a way to report phishing if you do identify/suspect foul play.

1. Be very careful when you click a link or open an attachment in an email. Even if an email appears to come from someone you know, that person's computer could have been compromised. If you see anything that looks suspicious (an obviously misspelled word or an email address you don't recognize), don't open any email attachments or click any links.

2. Type addresses yourself. Sometimes a link might appear to go somewhere safe. However, once you click the link it could send you somewhere else. If your bank or credit card company seems to be trying to reach you, go to your account the way you normally would (by typing their URL or clicking on a site you have bookmarked). Or pick up the phone and call them to be sure.

3. Turn on two-factor authentication. Some secure accounts have an option where you can ask for a code to be sent to your email or cell phone before you can access your account information. Sure, it's an extra step. But it could save you a lot of headaches in the long run.

4. Finally, you can forward suspicious phishing emails to spam@uce.gov or file a report with the Federal Trade Commission at FTC.gov/complaint.

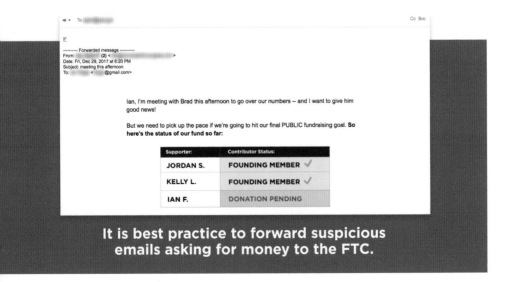

It is best practice to forward suspicious emails asking for money to the FTC.

WHAT ARE VIRUSES?

A virus is a type of malicious software program—"malware," for short—that can infect computers and replicate itself over and over, potentially harming several computers or entire networks of computers in a short period of time.

Without getting too technical, a virus needs a functioning program to "attach" itself to so it can do harm within the computer system. A virus can remain dormant within a computer until the program it has attached itself to runs, or executes. At that point, the virus can infect any number of functions within the computer or network, corrupting files, stealing IDs and passwords, or even sending spam emails from your account to those in your contact list, thereby potentially infecting their computers, too. There are few limits on the amount of damage a virus can inflict.

Many viruses find their way into computers through email attachments, so it's no secret that being vigilant when it comes to opening emails and attachments is one of the main ways to prevent infections to your computer.

Like a health virus (the flu, for example), a computer virus can spread from host to host. In fact, it is designed to do just that.

TYPES OF VIRUSES

Depending on who is counting, there may be as many as a dozen different types of viruses or as few as five. Some of the most dangerous are known as boot sector viruses, which infect computers at the boot, or start-up, level. This can damage the code operating the entire computer.

There are browser hijacking viruses that can route you to sites—sometimes dangerous ones—that you did not intend to visit. There are resident viruses that can run even after the original source of the infection has been eradicated. And there are file infector viruses that, as their name indicates, can cause havoc with files on your computer.

```
0 00 00-6D 73 62 6C                    msbl
0 6A 75-73 74 20 77  ast.exe I just w
9 20 4C-4F 56 45 20  ant to say LOVE
0 62 69-6C 6C 79 20  YOU SAN!! billy
0 64 6F-20 79 6F 75  gates why do you
3 20 70-6F 73 73 69   make this possi
0 20 6D-61 6B 69 6E  ble ? Stop makin
E 64 20-66 69 78 20  g money and fix
7 61 72-65 21 21 00  your software!!
0 00 00-7F 00 00 00    ♠ ♂♥►    H    △
0 00 00-01 00 01 00  ♂_♂_      ☺    ☺ ☺
0 00 00-00 00 00 46  ♠☺         L        F
C C9 11-9F E8 08 00       ♦]êèù└┌♦ƒ►☐
0 00 03-10 00 00 00  ♦►H `☐     ♠    ♥►
3 00 00-01 00 04 00  ▶♥   õ    ♂♥  ☺ ♦
```

A screen shot of the code for the computer-worm virus Blaster, which attacked Windowsupdate.com in the summer of 2003. Within the code, the hackers left a message for Microsoft founder and CEO, Bill Gates.

PREVENTING VIRUSES

Having up-to-date virus protection and being vigilant in your computer usage are key to keeping your computer and its files safe!

STEPS YOU CAN TAKE

The first step toward protecting your computer from a virus attack is to make sure you have an up-to-date anti-virus protection program installed and running on your machine. There are several to choose from, and annual subscriptions will only set you back perhaps $20 to $40. Some of the more popular ones are McAfee, Norton by Symantec, Kaspersky, and Bitdefender. There are others, too, and many computers come with a free trial of one of them.

When you install the anti-virus protection, make sure it runs by default every time you start up the machine. You should also make sure it is scanning your emails, in addition to the programs your computer runs.

As noted on the previous pages, viruses often spread via email attachments. The email might look like it came

Norton offers a variety of packages that can handle all types of digital environments.

from a friend of yours. If it was sent by a virus, though, the text in the body of the email might appear unusual or contain misspellings. If that's the case, or if you have any doubt, don't open it. You can have your anti-virus protection program scan it, or you can simply mark it as "spam."

By the same token, never download anything onto your computer unless you are certain it's coming from

a legitimate and trusted site. Finally, any time you plug an exterior device such as an MP3 music player or a storage device into your computer, be sure to run a virus scan.

It is important to note that both McAfee and Norton are subscription services that will require yearly payments to keep your system secure over time.

WHAT IS RANSOMWARE?

The fictional film *Alien* actually served as inspiration for a pair of Columbia University computer researchers who, in the 1990s, invented a method of encrypting computer data that makes the data unreadable to those without the decryption key.

The researchers had no way of knowing that their discovery would be used by cyber criminals to extort ransom money from innocent victims. Here's how it generally works: The criminal sends this form of malicious software—known as ransomware—into the victim's computer or computer network and encrypts an important set of data. That is, the data is rendered unreadable and unusable by the victim who owns the data. The data can either be locked in such a way that the victim can no longer access it without a code known only by the perpetrator, or

A screenshot of Petya ransomware. Petya was discovered in 2016, and although it did not infect as many computers as other ransomware in 2016, it was noted by IT security firm Check Point as "the next step in ransomware evolution."

altered in such a way that if the victim attempts to decrypt it, further damage is done to the data or the computer system.

The criminal now has the victim in a position to collect a "ransom." In other words, large sums of money are demanded from the victim in order to retrieve the data, or sometimes to prevent the publishing of the data if that information is sensitive or could potentially do harm.

Both governments and private businesses all over the world have been victimized by ransomware schemes over the last several years.

PREVENTING AND DETECTING RANSOMWARE

As dangerous as ransomware can be and as sophisticated as cyber criminals seem to become, the good news is there are some simple things you can do to protect yourself from being victimized by a ransomware attack.

STEPS YOU CAN TAKE

First, as we have covered earlier in this book, always make sure you have a reliable anti-virus protection program running on your computer. Such a program (like McAfee, Norton by Symantec,

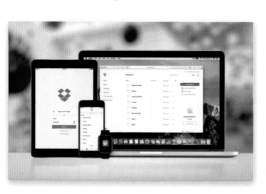

You can back up your files on the "cloud" by using an online service such as Dropbox or Google.

Kaspersky, etc.) can identify potentially dangerous emails or email attachments—the preferred ways of launching a ransomware attack.

Next, always be sure to back up your important data. In the business world, it's hugely important for companies handling sensitive data to back up that data on separate servers. That way, if the data is compromised, it can be restored from another source. Smart companies back up their data with a high degree of frequency. It's a good idea for individuals to back up their data, too.

Of course, being vigilant also helps prevent a ransomware attack, just as it does for any threat to your computer's security. Do not open suspicious attachments you might see in an email. Avoid clicking on links that seem to be questionable in any way. And never download a program or any information from a site that you don't know to be a reputable and legitimate one. A good rule to follow: "If in doubt, don't."

You can also back up your information by purchasing an external hard drive. You can attach the external hard drive to your computer and then detach it once data has been stored there.

WHAT IS CYBERFRAUD?

Any fraud committed using the internet can be called cyberfraud or internet fraud. Of course, this is a wide-ranging definition. There are, however, certain types of cyberfraud that tend to be the most popular online crimes. Over the next few pages, we will point out some of the more common methods and discuss ways of protecting yourself against cyberfraud.

TYPES OF CYBERFRAUD

Consumer Fraud: Automobile buyers are a common target for criminals looking to make a quick buck illegally. The criminals might post a listing for a vehicle that doesn't exist, luring unsuspecting consumers. When a potential consumer expresses interest, the "seller" might tell them that the vehicle is located out of the country, and request that the consumer transfer funds to a third party to "hold" the vehicle and protect the purchase and/or shipping of the vehicle. Funds then get wired somewhere the criminal can access them, and only after that might the consumer realize there is no car.

Consumer fraud is not limited to the auto industry. With more and more people willing to make purchases over the internet, there has also been a rise in those trying to swindle online consumers. If you are making a large purchase, make sure you are doing so through a reputable online retailer that has a consumer protection policy. Knowledge is power.

Charity Fraud: Just as criminals will prey upon consumers looking to make online purchases, they might also decide to profit from the generosity of those who give back. This is especially rampant after deadly or catastrophic natural disasters or weather events, when there is an outpouring of public generosity. Scammers have set up fake charities to turn those "donations" into profit.

When making a donation for any charitable cause, either stick with a known agency that you know to be "above the board" such as the American Red Cross, or do some research to make sure the hard-earned money you are generously giving goes to help the intended recipients.

TIPS TO PREVENT CHARITY FRAUD

- Do not respond to unsolicited emails
- Be skeptical of individuals claiming to be officials via email or phone
- Do not click links or attachments in unsolicited emails
- Do not rely on others to donate on your behalf
- Authenticate the legitimacy of the charity by contacting them directly
- Do not provide personal or financial information to solicitors

TYPES OF CYBERFRAUD

Ticket Fraud: Just as counterfeit tickets are routinely sold at sports, music, and other venues around the country, they are also big business in the world of cyberfraud. You may think you are buying floor-level seats to see Adele when she comes to town. You might actually be purchasing a bogus and worthless piece of paper that won't get you in the door. Only purchase online tickets from a known and reputable ticket sale site such as Ticketmaster or StubHub. Even though some sales on those sites are actually resold tickets from other customers, they come with a consumer protection guarantee.

Online Gambling Fraud: Offshore companies that have not been beholden to American law have been competing for U.S. gambling dollars for years. Some have been

known to pocket your winnings. Others have "gone under" as businesses without reimbursing owed funds. The best advice is to steer clear of online gambling.

Ticket fraud is so prevalent that it has even impacted worldwide attractions like Super Bowl LII in Minneapolis in 2018 and the Rio Olympics in 2016.

Online gambling provided a $645 million revenue for gambling sites in 2015, and the market is expected to grow in the U.S. each year by fifty percent until 2020. This market is a humungous potential cash cow for fraudsters.

Gift Card Fraud: At some point, you have probably held a store gift card that had a remaining balance. That is, you spent $35 of the original $50 card value and have $15 left on the card.

Don't put it past cyber criminals to go after that $15. Programs have been written to track down gift card numbers from stores of all kinds. Once the scammers have those numbers, they can use the store websites to check the balance on those cards. That cash value can either be used or re-sold to others, putting money in the criminals' pockets.

This is a tricky one, but using the full amount of a gift card not long after you receive it is probably the easiest way to avoid this type of cyberfraud. You can also check the remaining value yourself. If a card that had a remaining balance now shows no such value, contact customer service at the retailer immediately.

THE FEDERAL TRADE COMMISSION HOTLINE

1-877-FTC-HELP

Or for more information, go to
www.consumer.ftc.gov or
www.ftc.gov/subscribe

WHAT IS MALWARE?

Malware is a broad term used to classify a variety of malicious and intrusive computer software from viruses, worms, Trojan horses, ransomware, spyware, adware, and scareware.

WHAT IS MALWAREBYTES?

First released in 2006, Malwarebytes is anti-malware software that helps keep your computer and computer network safe. It goes a step further than some other anti-virus software in that its scans and removes several types of malware, spyware, and adware that are sometimes elusive to other methods.

An update to the software (3.0 version) that was released in December 2016 included protection against ransomware. Malwarebytes has been generally well received by computer publications and is worth considering when deciding how to secure your computer.

SELECTING WHAT YOU NEED

There is a free version and two paid versions of Malwarebytes. The free version complements the anti-virus program you (hopefully) already have running on your computer and is a great idea to download and install. As of this writing, the free version works to repair infected programs and devices only on Windows computers.

The lower-cost paid version can replace your current anti-virus program, as it will take care of all the same scanning functions in addition to repairing infected items. That software works on Windows, Macintosh, and Android devices.

There is also a higher-cost version marketed toward small businesses. It comes with a higher number of devices it can protect and includes a priority level of customer support businesses might require. Businesses can sign up for a free Malwarebytes trial before deciding whether to purchase the software.

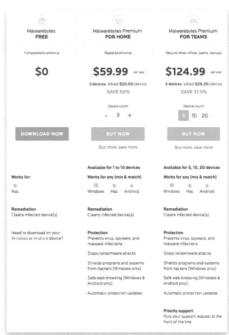

USING MALWAREBYTES

To download the free version of Malwarebytes, visit Malwarebytes.com and click the **Free Download** button. Choose your language and then click **Agree and Install**.

Once the program opens, simply click **Scan Now** to set it to work scanning your computer for any threats. While the premium (paid) version of Malwarebytes is constantly scanning your computer for potential threats, you will need to run the free version manually. Fortunately, it's as easy as the click of a button.

With Malwarebytes Premium, you have advanced options for scanning your devices. A Custom Scan allows your to scan specific areas of your device's operating system without scouring the whole system.

Hopefully, you will see that no threats have been detected after a scan that should not take more than a few minutes. If there are any threats detected, Malwarebytes will quarantine the questionable files to keep them from doing damage (or further damage, if any has been done).

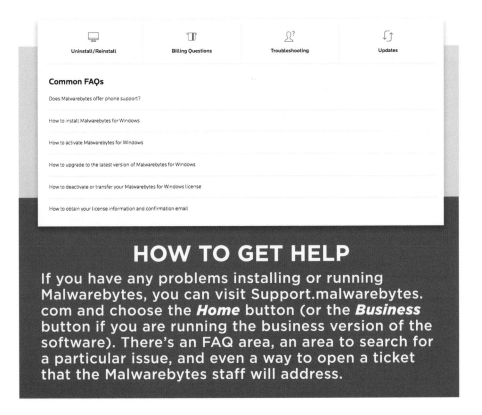

HOW TO GET HELP

If you have any problems installing or running Malwarebytes, you can visit Support.malwarebytes. com and choose the *Home* button (or the *Business* button if you are running the business version of the software). There's an FAQ area, an area to search for a particular issue, and even a way to open a ticket that the Malwarebytes staff will address.

WHAT IS IDENTITY THEFT?

A study by Javelin Strategy & Research found that $16 billion was stolen from more than 15 million Americans in 2016 through identity theft. Every few seconds, the study found, someone's identity is stolen.

We have seen many different types of cyber fraud and internet scams outlined in the previous pages, and learned several ways to protect yourself and your computer from becoming victimized. Now let's have a look at how you can minimize your chances to have your very identity breached.

HOW YOUR IDENTITY IS STOLEN

There are many ways criminals can steal your credit card information. Some victims have had their card numbers stolen at ATM machines by having long-range photographs taken of either their cards, their PIN entries, or both. Criminals sometimes sneak small devices that

26

You can ask for a credit freeze through the three main American credit bureaus (Equifax, Experian, and TransUnion).

can steal information into card readers at places like gas pumps and convenience stores.

If you notice false charges on your credit card account or are notified by your card company or bank that there is suspected fraudulent activity, your credit can be temporarily frozen, new cards issued, and fraudulent charges removed.

PREVENTING IDENTITY THEFT

There are services you can hire to add various levels of credit monitoring that can help protect your identity from being stolen. One of the most popular is LifeLock. The Arizona-based company actively searches for threats to its subscribers' identities, notifies subscribers when threats are detected, and in the case of stolen identity or fraudulent activity, works with the subscriber to restore funds and safety.

Criminals steal identities in order to take advantage of their victim's identity to sign up for credit cards, get out of a crime, obtain medical care, or even assume their victim's identity on a day-to-day basis.

USING LIFELOCK

In a world full of schemes and individuals running hi-tech scams, it can be very difficult keeping your money and your identity safe. LifeLock is a service that can help you stay secure.

SIGNING UP FOR LIFELOCK

One can sign up for LifeLock for as little as $10 per month. Simply visit Lifelock.com and click on the membership of your choice. The most affordable membership provides social security number and credit alerts, and will reimburse up to $25,000 if funds happen to be stolen. For double that cost (as of this publication), that amount increases to $100,000 and additional protection is provided for bank and credit card activity, along with alerts on any crimes committed in your name. There is also a premium option that provides $1 million in reimbursement.

No matter which one you select, be sure to be thorough in providing the information the service needs to keep your identity safe. There is also a phone number right at the top of their home page if you feel more comfortable talking to a representative.

LIFELOCK'S SERVICES

The services offered include lost wallet protection, verification of address changes, data breach notifications, and "dark web monitoring." Without getting too technical in these pages, search engines and "regular" internet use only access parts of the web that are made accessible to such searches and the web browsers you use. There is a world of information accessible only through specialized programs, or "darknets." Here, peer-to-peer communications are not indexed for typical web use.

How LifeLock works to help protect you from identity theft.

SIGN UP
① It only takes a few minutes to enroll.

WE SCAN
② We look for threats to your identity.

WE ALERT†
③ We alert you of suspicious threats by text, email, or phone."

WE RESTORE
④ If your identity is stolen, a U.S.-based Identity Restoration Specialist will work to fix it.

WE REIMBURSE
⑤ We'll reimburse funds stolen due to identity theft up to the limit of your plan.

$300.00
The Carphone Warehouse

Did you make this transaction?

YES   NO

Alert Date   March 02
Alert Type   Purchase
Data Source   Bank
Alert Category Credit Card - Retail
Activity Type   Regular

It can be a rogue world where information is traded on the dark web where it is more difficult to monitor. LifeLock can help bridge that gap.

EQUIFAX, EXPERIAN, AND TRANSUNION

You might have heard about Equifax because of 2017 news reports. The company, a consumer credit reporting agency that monitors information on more than 800 million consumers worldwide, experienced a cyber security breach not unlike the ones we covered earlier in these pages. Many people's information was compromised.

As one of the three largest credit monitoring agencies— along with Experian and TransUnion—this was major news, and a major problem. More than 200,000 credit card credentials were stolen by cyber thieves, and more than 145 million Americans (almost half the population) were impacted by having information such as drivers' licenses, addresses, phone numbers, and social security numbers stolen.

The company waived fees for security freezes in the immediate wake of the breach, and offered their Lock & Alert services for free to those who signed up before January 31, 2018.

SIGNING UP FOR FREE CREDIT MONITORING

To sign up for Equifax's Lock & Alert service, visit Equifax.com. You can also type in your social security number to see if your information was, indeed, compromised in 2017 and get information about what next steps might be smart in making sure your accounts and finances have not been victimized by theft.